C. G. C

# THE SPECIAL AND GENERAL THEORY OF RELATIVITY

*Simplified Explanation*

*The Special and General Theory of Relativity: Simplified Explanation*



ISBN-13: 978-1545005101
ISBN-10: 1545005109

1. Science / General
2. Science / Physics

# Introduction

In this short book we shall go through—with a simplified approach—the logical sequence of deductions that let Albert Einstein formulate his great relativistic theories, that were extremely revolutionary in the field of Theoretical Physics, but also paramount for a more correct, and complete, comprehension of the phenomena that take place in the universe.

This treatise uses very simple words. By purpose the author has "narrated" the subject as much as possible avoiding difficult mathematics. Where necessary, the physical-mathematical passages has been described by means of formulas that should be very easy to understand.

In order to fully understand and appreciate this book, it is then sufficient to have scholastic basics correspondent, approximately, to the secondary school. Some particular concepts, a bit more complex, will be in any case clearly explained in this book.

In fact, the aim of this essay is to describe the main points of the Special and General Theory of Relativity in a lean and possibly enjoyable way, without demanding to the reader a particular preparation in Physics or in Mathematics.

So, you will not find here a cold treatise on Relativity. Experts in physics may realize that some details, or evidences, have been left aside. But this is, again, explicitly by purpose.

Let us begin with the first theory, named Special Relativity, formulated by Einstein in 1905. Then we will proceed with the general theory, called in fact General Relativity, published in 1915.

# THE SPECIAL THEORY OF RELATIVITY

## Basic hypothesis

The Special Theory of Relativity is valid under some particular conditions.

The main hypothesis is that the reference frames that we shall consider must be "inertial:" that is to say that this theory does not apply to all those cases in which there is an acceleration of some sort (linear, centrifugal, etc.).

The velocities "in the game" must therefore be *constant* (or zero).

The non-inertial systems—the ones that have an acceleration—will be taken into consideration in the following relativistic theory developed by Albert Einstein: the General Theory of Relativity. That theory takes into account the more general and complete situation, normal in the physical reality in which we live.

However, even though it is based on the behavior of physical entities in the very particular case of absence of both accelerations and gravitational fields, the Special Theory of Relativity reaches so *extraordinary* conclusions regarding the way the universe works, that it has opened the human mind to a very profound comprehension of the physical reality, and it has also led to theoretical and practical applications of great importance.

# A first, important step

Let us consider two persons that are traveling on two different starships, and such space vehicles are getting away the one from the other with a uniform, straight motion at constant velocity.

*Starship A* *Starship B*

Be $v$ the velocity of the starship $A$, and $w$ the velocity of the starship $B$.

The person in $A$, if he does not have any points of reference other than the possibility to look at the starship $B$, can think, without being wrong, that he is resting in space, while the spaceship $B$ is flying away at a certain velocity. A velocity that, if calculated, would be equal to $v + w$.

At the same time, the person in $B$ could think to be remaining still in space, while looking at the starship $A$ going away at a velocity $v + w$ in magnitude.

Now let us add a third reference frame: ours. Because we consider ourselves "still" by hypothesis, in respect to *A* and *B*, so that we see the starship A flying away to the left with velocity *v* and the starship *B* to the right with velocity *w*, the statement "we are resting, and the starships *A* and *B* are flying away from us" is true as well.

No experiment can be done by the astronaut in *A*, or by the one in *B*, to realize to be in a state of motion, in an "absolute" sense.

# The Michelson-Morley experiment

One important fact, on which the Special Theory of Relativity is based, is unequivocally the experiment performed by physicists Michelson and Morley, in their attempt to measure the speed of the Earth in space.

They knew that the Earth is moving around the Sun, and that the Solar System is orbiting the center of our galaxy, the Milky Way, but they wanted to measure the velocity of the Earth in an absolute sense, that is to say *in respect to* space itself.

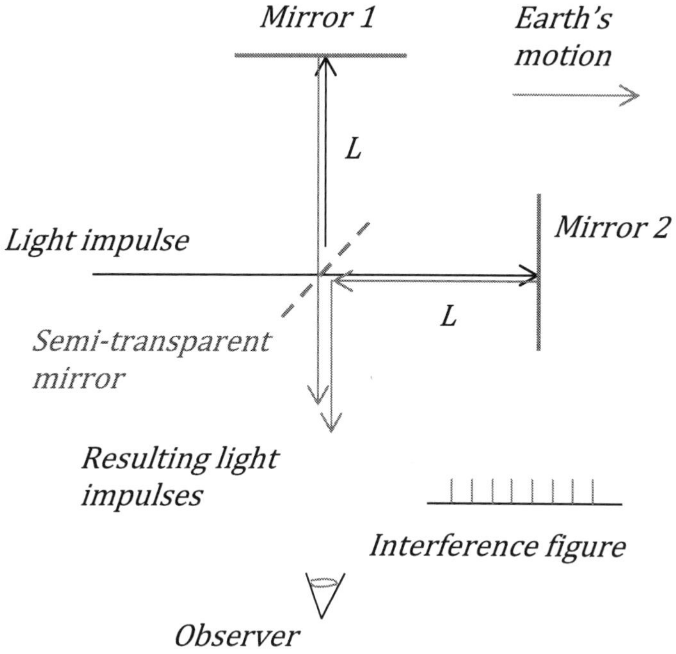

The picture here above summarizes the experiment done by the two scientists.

One single light impulse is emitted in the direction of a mirror that reflects light photons only partially ("semi-transparent mirror" in the picture). So the light is partially reflected toward the mirror number *1*, placed at a distance $L$ from the semi-transparent mirror, and partially it passes through the semi-transparent mirror itself, continuing its travel toward the mirror number *2*, that is also placed at a distance equal to $L$.

The two consequent reflections (I have drawn them not overlapped to the incident rays only for clearness) travel back to the semi-transparent mirror, and of those two, we consider only the path that in the picture goes toward the observer.

As a measurement term we take the *interference figure* that results from the sum of the two light impulses.

If the Earth were in motion in respect to space in an absolute sense, then by rotating the experiment around its center (point in which the beam passes through or is reflected by the semi-transparent mirror) we should see different interference figures, because the light would take less or more time to travel the two intervals of space with length L, depending on if or not (an to what extent) it should go "countercurrent" in respect to the space that "flows underneath" the Earth.

But with great surprise of the two experimenters, no difference at all was measured.

Only two are the conclusions of Michelson and Morley's test: either the Earth is motionless in respect to space, or the speed of light is invariant if measured from different *inertial* frames of reference.

But the first hypothesis is clearly wrong, because we know for sure that the Earth orbits the Sun, that the Solar Systems rotates around the Milky Way's center, and so on. That is to say the Earth is not at rest.

Thus, *the speed of light is independent from the speed of the observer, and also from the speed of the source that emits the light itself.*

This result is unquestionably counter-intuitive, but it is a measurable physical reality, and upon it an important part of Einstein's reasoning relies. A reasoning that—we shall see—brought to the Special Theory of Relativity.

# The mental experiment of the train

We now want to repeat one of Einstein's "mental experiments."

So we consider ourselves on a train that is traveling at a constant velocity $v$ , equal to *50 miles per hour.*

Let us throw a ball in the same direction of motion of the train, at a speed $w$ = *10 miles per hour:*

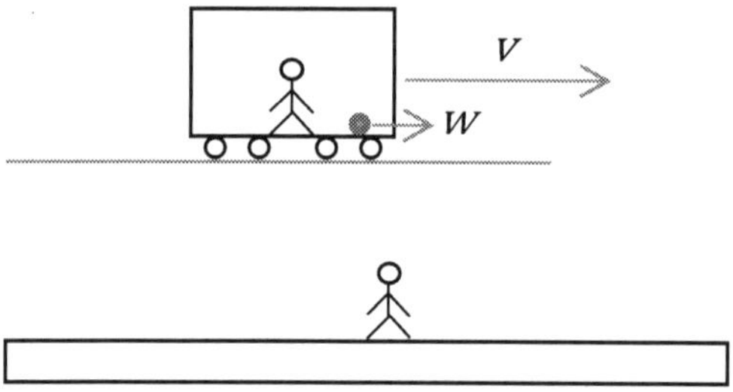


$w$ is then the relative speed of the ball in respect to us, because we are standing on the train.

An observer, standing still on the railway platform, sees the train traveling at *50 miles per hour,* and so he sees the ball running at *60 miles per hour* in respect to him, because:

$$v + w = 50 Km/h + 10 Km/h = 60 Km/h$$

So far, so good.

Let us now do again the same mental experiment, but this time we substitute the ball with a light beamer (drawn as a small triangle, in the picture here below), that emits a luminous impulse.

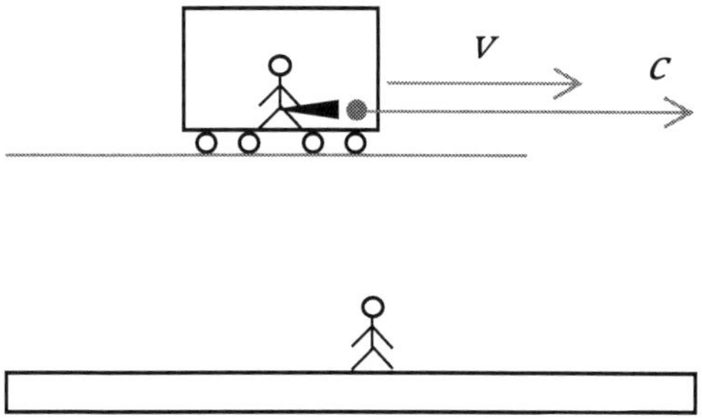


We would expect that the observer on the railway platform sees the light impulse traveling at the velocity:

$$v + c = 50\ Km/h + 3*10^8\ m/s^2$$

But that does not happen.

In fact, thanks to the Michelson-Morley experiment, we know that the observer would measure a propagation speed of the light impulse exactly equal to $c$, even though he is standing still in respect to the train.

That is to say that, both in the case we travel on the train at a constant speed, and in the case we are standing still on the railway platform, we would measure exactly the same speed of the light impulse, emitted from a light beamer placed on a moving train.

This strange fact, that with no doubt is counter-intuitive, has some very peculiar physical consequences, as we are going to see.

# The relativistic correction and the Lorentz factor

Let us now consider this situation, in which there are again two reference frames, one "at rest" and the other in a relative motion in respect to the first one, with a constant velocity.

Be the first reference frame defined by the space-time variables $x$ and $t$. And, in the second reference frame, $x'$ and $t'$ be, respectively, the correspondent variables of space and time.

That is:

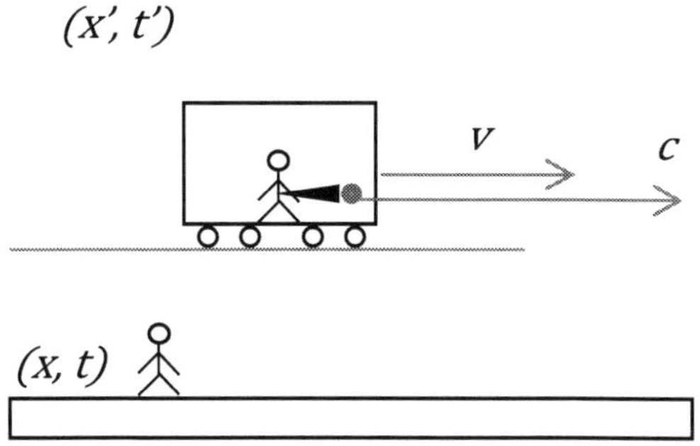


On the space-time graphic that we can draw for the observer (measured in seconds and meters) we plot the velocity of the train $v$ and the velocity of light $c$ :

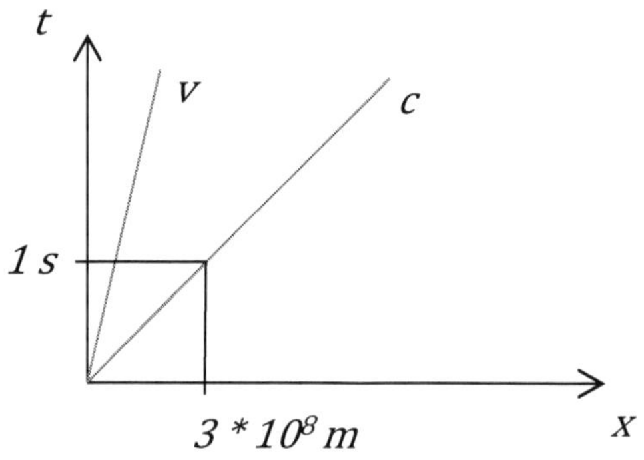


The velocity of the train, measured by an observer on the railway platform is equal to:

$$v = \frac{x}{t}$$

because that speed is equal to the space $x$ traveled by the train, divided by the time $t$ that elapses in the movement. This is true just by definition of velocity.

Equivalently:

$$x = v \; t$$

or:

$$x - v \; t = 0$$

Let us now consider the reference frame $(x', t')$, bound to the train.

A person that is traveling on the train does not measure any motion in respect to the train itself. Thus:

$$x' = 0$$

By matching the two previous, very simple formulas, we obtain:

*(1)* $$x' = x - v\ t$$

For the moment we assume that the time measured in the two reference frame is equal, that is to say that we assume:

$$t' = t$$

a fact that, we shall see, in reality it is not correct.

Now, if we measure the speed of light $c$, in respect to the two reference frame, we get:

$$c = \frac{x}{t}$$

or, equivalently:

$$x = c\ t$$

and:

$$x' = c\ t'$$

with $c$ being the same in both cases, thanks to what we learned from Michelson and Morley's experiment.

By substituting in the *(1)* we get:

$$c t' = c\ t - v\ t$$

If the condition $t' = t$ were really true, by dividing all by $t$ we would get:

$$c = c - v$$

that is absurd, because $v$ is not equal to zero.

Thus, there is a mistake somewhere, and now we want to find it, and quantifying it.

Let us take again the *(1)*, that we can also write as:

$$x = x' + v\ t$$

and let us substitute $t$ with $t'$, because we are assuming $t = t'$. We obtain:

*(2)* $$x = x' + v\ t'$$

we can think of the formulas *(1)* and *(2)* as the measures of space in respect to the two reference frames.

But, as we saw, something was wrong with our reasoning, and then we introduce, with the aim to identify it, an error term that we define $\gamma$ ("gamma"):

$$(1a) \qquad x' = \gamma \ (x - v \ t)$$

$$(1b) \qquad x = \gamma \ (x' + v \ t')$$

By multiplying the corresponding terms we have:

$$x' x = \gamma^2 \ (x - v \ t) \ (x' + v \ t')$$

and so:

$$x' x = \gamma^2 \ (x x' + x \ v t' - x' v \ t - v^2 \ t t')$$

but because we know that $x = c\ t$ and $x' = c\ t'$, we get:

$$t = \frac{x}{c} \quad \text{and} \quad t' = \frac{x'}{c}$$

thus, by substituting in the above written formula, we obtain:

$$x' x = \gamma^2 \ \left( x' x + \frac{x \ v x'}{c} - \frac{x' \ v \ x}{c} - \frac{v^2 \ x x'}{c \ c} \right)$$

and, dividing all by $x' x$:

$$1 = \gamma^2 \ \left(1 + \frac{v}{c} - \frac{v}{c} - \frac{v^2}{c^2}\right)$$

As you can see, two terms cancel out and we have:

$$1 = \gamma^2 \left(1 - \frac{v^2}{c^2}\right)$$

Now, by applying the square root on both of the terms we get:

$$1 = \gamma \sqrt{1 - \frac{v^2}{c^2}}$$

or, equivalently:

$$\gamma = \frac{1}{\sqrt{1 - \frac{v^2}{c^2}}}$$

Thus, finally we identified the error $\gamma$.

$\gamma$ is called Lorentz factor, or Lorentz transformation term.

If we substitute such a factor in the *(1a)* we have:

$$x' = \frac{x - vt}{\sqrt{1 - \frac{v^2}{c^2}}} \qquad (3)$$

This formula represents, de facto, the "relativistic correction" on space, in the case in which the velocity $v$ is not neglectable in respect to the speed of light $c$.

Please note that in the opposite case, that is to say $v \ll c$, the denominator tends to the value *1* and we have again the formula *(1a)*.

The same reasoning is obviously applicable to the formula *(1b)*.

Let us recap.

If we speak about velocities *far lower than the speed of light*, like the velocity of a train, for instance, all what we have to do is to apply the formulas *without* the relativistic correction. Or, equivalently, if we apply the formulas with the relativistic correction, the term in the denominator tends to 1.

But if the velocities involved are comparable to the speed of light, then we have to take into account the relativistic correction, and we have to use the Lorentz factor: the term in the denominator is substantially smaller than 1.

# Consequences on space

Let us write again the formula that we obtained in the previous chapter, and let us reason on its physical consequences:

$$x' = \frac{x - v\,t}{\sqrt{1 - \frac{v^2}{c^2}}} \qquad (3)$$

For velocities *v*, comparable to the speed of light, the term in the denominator tends to *0*, because it is equal to the square root of a term that becomes *1 minus circa 1*. While, for velocities *v* far lower than the speed of light, the denominator tends to 1, because it becomes equal to *1 minus circa 0*.

With this in mind, let us try and imagine we want to measure the length of an object, for instance a little rod, that is on a train, running with a constant speed:

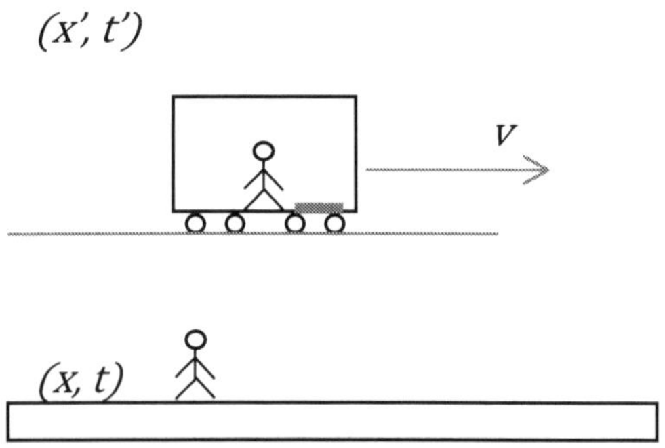

We want to measure that rod both in respect to the reference frame bound to the train *(x', t')*, and in respect to the reference frame of the railway platform *(x, t)*.

In order to do this, we have to determine the coordinates of the points at the extremes of the rod. And if we want to measure the length of the rod, we do not want time to pass, between one measurement and the following one. Otherwise we get a surely wrong length in the measurement made from the railway platform, because of the fact that the train is moving.

Thus, we have to measure simultaneously the two extremes of the rod. That is as much as to say that $t = 0$.

By imposing that $t = 0$, the previous formula simplifies into:

$$x' = \frac{x}{\sqrt{1 - \frac{v^2}{c^2}}}$$

that we can also write as:

$$x = x'\sqrt{1 - \frac{v^2}{c^2}} \qquad (4)$$

Now, because the term under the square root has values between *0* and *1,* we obtain a first result, as much important as odd: every physical object, placed on a vehicle that travels at a speed comparable to the velocity of light, appears *contracted* in the direction of motion, if its length is measured from a reference frame "in rest," in respect to such a vehicle.

That is to say, that object appears shorter in respect to the measurement we can do by standing on the moving vehicle.

# Consequences on the flowing of time

Let us now reason about time.

Let us consider a formula analogous to the *(3)*, that we obtain by substituting $\gamma$ in the *(1b)*:

$$x = \frac{x' + v t'}{\sqrt{1 - \frac{v^2}{c^2}}}$$

This time we do not impose *t' = 0*, just because we want to see what happens to the flowing of time.

Let us consider an experiment in which light is involved. Be *x'* the space that light travels, in respect to the reference frame bound to the train, during the amount of time *t'*. Thus:

$$c = \frac{x}{t} = \frac{x'}{t'}$$

Or:

$x' = c\,t'$; $x = c\ t$ and also $t' = \frac{x'}{c}$

Simply by substituting these values in the formula we wrote in the beginning of this chapter, we get:

$$c\, t = \frac{c t' + v \frac{x'}{c}}{\sqrt{1 - \frac{v^2}{c^2}}}$$

Dividing by $c$ we have:

$$t = \frac{t' + v \frac{x'}{c^2}}{\sqrt{1 - \frac{v^2}{c^2}}} \qquad (5)$$

Thus:

$$t \neq t'$$

***t'*** is the time measured on the train and ***t*** is the time measured on the railway platform.

If the train travels at a velocity very lower than the speed of light, the terms $\frac{v}{c^2}$ and $\frac{v^2}{c^2}$ both tend to zero, and so we have the particular case, normal for us, of $t = t'$.

If now, for simplicity of explanation, we consider an event that marks the time flowing on the train, and that does not occupy any *space* on the train (for example a clock of infinitesimal dimensions, or only its "tick-tock"), we have, by definition, $x' = 0$ in the formula number *(5)*.

Thus, we have:

*(6)*

$$t = \frac{t'}{\sqrt{1 - \frac{v^2}{c^2}}}$$

from which it is evident how the apparent flowing of time differs in the two situations (platform and moving train): measured from the railway platform time $t'$ dilates (it flows slower) as we approach the speed of light.

This means that the clock of a person that is traveling on the train is seen marking the time slower than the clock of a person that is standing still on the railway platform.

Please note that, if we wanted to use the formula *(3)* of the previous chapter, and we substituted in it the values $x' = ct$; $x = ct$ and $t = \frac{x}{c}$, we would have obtained a formula *similar* to the number *(5)*. But then we could have not impose the condition $x = 0$, in order to reason on the time only, because $x = 0$, very differently from $x' = 0$, means that the train is *at rest*, if time passes.

But let us come back to our considerations.

Summing up, we can now put together the two conclusions about the flowing of time and about the deformation of space.

To visualize the two concepts together, we can think about a clock with certain dimensions, moving bound to a very fast train. From the railway platform we would see the clock *contracted* in the direction of motion and its marking of time (for example the movement of its hands) *running slower*.

In order to better understand this extraordinary relativistic effect, we can try and calculate the contraction of a clock diameter after the formula *(4)* of the previous chapter, and the slowing of the clock hands after the *(6)*.

For instance, we consider the situation in which the velocity of the train is equal to *0.9 c* (90% the speed of light), and we assume the clock has a diameter of *1 meter* , if measured in the reference frame bound to the train.

We omit the mathematical passages only for synthetize, but it is easily demonstrable that the clock, in the reference frame *(x, t)*, appears contracted, and shows a "diameter," in the direction of the motion, equal to approximately *0.44 meters*. And the time flowing that is marks is *2.29 times slower,* in the sense that its "seconds" (intervals of time between a "tick" and a "tock") last *2.29 seconds*, if measured with another clock that is standing still on the railway platform.

# The twin paradox

From the considerations of the previous chapters, and in particular from the effect of time dilation, that occurs when the considered velocities are comparable to the speed of light, the famous example, called "the twin paradox," derives; here we want to explain it very shortly.

First of all, let us bear in mind that what happens to the clock that we just considered obviously happens to all the *clocks* that are present on the train, biological "clocks" —the ones in a passenger's body, for instance— included.

Thus, a first deduction is that you cannot notice any difference in time flowing, if you stand still on the train, if you do not have any other references in addition to the environment in which you find yourself (the train itself).

And so, for a person that travels on the train, time is flowing quite normally: he cannot notice any slowing in time flowing caused by the velocity of the train, because he, as an observer, is slowed in his biological and cerebral functions to the same extent. And the same happens to every measurement device that may travels on the train: if you use such instruments to measure time or space on the train, you cannot notice any difference even at velocities so high to be comparable to the speed of light.

Let us now consider two twin brothers.

One of them begins a space journey on a starship that can travel close to the speed of light.

As measured by the brother that remains on the Earth, the journey lasts a certain amount of time. Let us say 20 years.

At the end of the trip, for the twin brother that remained on the Earth a lot of time passed, and he appears, obviously, exactly 20 years older.

But with great surprise, when the starship hatch opens, he sees his brother evidently much younger than him. In fact, his twin brother appears to have aged, for instance, a couple of years.

What happened?

Thanks to the formulas we derived until now, for us it is quite clear: by traveling on the starship at velocities close to the speed of light, for the astronaut *time dilated*, and so his biological processes, among which the process of aging, flowed much slower than the ones of his twin brother that remained on the Earth.

A concept to have clear in mind is that the twin brother that traveled in space has "earned" not even a second of life more than his brother who remained on the Earth. Simply, until his starship touched down on the Earth, he made a certain amount of actions (physical, mental, biological, etc.) that can be made in a couple of years of life, while his non-astronaut brother has done a number of actions correspondent to 20 years of life.

In other words, at the moment of the starship arrival, the astronaut has lived a couple of years of his lifespan, and his brother has lived 20 years of his lifespan.

And so, for the twin brother that traveled in space, 18 years more remain to be lived (he has not yet "consumed" them) in respect to his brother. This is true, obviously, if their life expectancy is exactly the same. But for the moment he has also lived much less, in the sense that he has done much less things in respect to his brother.

And if their life expectancy is exactly the same, then the two twin brothers will have lived, at the end, exactly the same number of years. Only, they will have lived such years in a *shifted way* one in respect to the other, due to the space trip—at velocities close to the speed of light—that one brother has taken.

# A real measurement that verifies time dilation

Did anybody ever verify, experimentally, the amazing phenomenon of time dilation, or is it only an extravagant theory?

The answer is yes, it was actually measured in more than a type of experiments.

Let us take into consideration the case of the muons.

The muons are particles that move at a velocity equal to 99% the speed of light. That is to say, $v = 0.99\ c$.

In addition, they have a half-life of $1.5\ \mu s$ (microseconds). This means that every $1.5\ \mu s$ their number (quantity) decreases by a factor of *2*.

In one experiment, physicists considered the muons originated by cosmic rays that hit the upper atmosphere of the Earth.

The muon flow was measured at the top of a mountain *2000* meters high, and then at sea level.

At the altitude of *2000* meters they measure a quantity of *520* muons per minute, that pass through a determined section.

At sea-level, through exactly the same section, they measured *290* muons per minute.

Because the difference in altitude, between the two places where the measurements were performed, was equal to 2000 meters, the transit time between such altitudes, for objects that travels at *0.99 c*, is of circa *6.73 μs*, as we could simply calculate by dividing the space to be travelled by the velocity.

But if the half-life of muons is equal to *1.5 μs*, then in *6.73 μs* the muons should have had the time to halve their number more than *4 times*. That is to say, they should have measured, at sea-level, only some *thirty* muons per minute:

*(((520 / 2) /2 ) / 2) / 2 = circa 30 per minute*

But they actually measured a much larger number: *290* per minute.

This fact suggests that for the muons time has flown slower, like the Theory of Special Relativity affirms.

Not only: by applying the formulas of Einstein's theory, we obtain a *calculated* dilation of time perfectly correspondent to what we can *experimentally* verify.

In fact, by performing the calculations, we get that for the muons it is just like only *0.94 μs* passed, not *6.73 μs*.

*0.94 μs* do correspond to less than the half-life of muons and in fact, experimentally, at sea-level they measured *290* muons per minute, that is to say more than the half of the *520* muons measured at an altitude of *2000* meters.

# What happens to mass?

We understood what happens to the flowing of time, and to space, when we move at so high a velocity that we almost reach the speed of light.

Now we think about the relativist effects on mass.

Let us put aside any physical demonstration, in order to remain on a simple level of narration.

The following one is the formula that we would obtain for the mass:

$$m = \frac{m'}{\sqrt{1 - \frac{v^2}{c^2}}} \qquad (7)$$

being *m* the mass measured from a reference system "at rest," and *m'* the mass of the same object, measured in the reference system in motion with the mass itself.

That is equivalent to say that the more the velocity *v* of a body of mass *m'* gets near to the speed of light *c*, the more the measurement *m* of the mass of the same body is larger than *m'*. In fact, the denominator of the above written formula tends to *0*.

This also explains why an object with a certain mass can never reach—let alone surpass—the speed of light: its mass would become infinite, and so it would be necessary an infinite force $F$ to accelerate it to such a velocity, because of the formula:

$$F = m \quad a$$

where $m$ is the mass of the object and $a$ the acceleration.

By the way, the photons of light do not have mass, and so they can travel, without contradicting any physical law, exactly at the speed of light.

Another consideration: when we talk about the speed of light—that as we saw is a universal constant, but also a universal limit to the velocities of objects—in reality we are not only referring to visible light, but to *photons in general*, and so to all the electromagnetic radiations, such as radio waves, microwaves, gamma rays, etc.

Even more, with the term "speed of light" we actually mean the speed of every particle that has *no mass*.

Photons are an example of "particle" without mass, but another example (the only one known today) is the gluon, that is the particle responsible for the strong nuclear force. Hence it has no mass, it could also travel at a velocity equal to $c$ (at least theoretically, but not experimentally, because we never observed a "free" gluon, i.e. outside the particles composed of quarks).

# $E = m c^2$

Let us now continue our exploration of the consequences of the Special Theory of Relativity.

By recalling into mind a mathematical notion, we know that:

$$(1+x)^n = 1 + n\,x + \ldots$$

In place of the dots, from a strictly *mathematical* point of view, there would be other terms, of a lower order, but for the *physical* considerations that follow, such terms are neglectable.

By using this notion in the formula *(7)* of the previous paragraph, that we now write in this equivalent way:

$$m = m'\,\left(1 - \frac{v^2}{c^2}\right)^{-1/2}$$

we obtain:

$$m = m'\,\left(1 + \frac{v^2}{2c^2} + \ldots\right)$$

and omitting, because of their irrelevance from a physical point of view, the additional terms that above we have substituted with some dots, we get:

$$m = m' + m'\frac{v^2}{2c^2}$$

finally, by multiplying by $c^2$ we have:

$$m\,c^2 = m'\,c^2 + m'\frac{v^2}{2}$$

The component $m'\frac{v^2}{2}$ is nothing but the *kinetic energy* possessed by the mass $m'$, that is travelling at the velocity $v$.

Then the second term of the above written equation is an energy. But then, also the first term must be an energy. So we can write:

$$E = m'\,c^2 + m'\frac{v^2}{2}$$

where $E$ stands for "energy."

$m'c^2$ is also defined as the "energy at rest" (i.e. at zero velocity).

If now we consider a mass actually at rest, we can omit the apices on the mass $m$ and the kinetic energy becomes zero. So it is evident how Einstein deduced his famous formula:

$$\boxed{E = m\,c^2}$$

for a whatever physical entity at rest.

This is, without any doubt, the most famous physical formula in the world.

# Let us recap...

By reading again the sequence of the main points of this short treatise, that tries to summarize in a simple, but correct way the long and revolutionary path walked by the great Albert Einstein, we can now do a "summary of the summaries" of the Special Theory of Relativity. In order to do that, the best thing is to list the 4 fundamental formulas:

$$t = \frac{t'}{\sqrt{1 - \frac{v^2}{c^2}}}$$

that tells us that the time $t'$ of a reference frame $(x', t')$, moving with a uniform linear motion at a velocity $v$ in respect to a second system of reference $(x, t)$, appears to be *dilated* if measured in $(x, t)$; in other words, it appears to flow slower, if measured in the second reference system $(x, t)$.

$$x = x'\sqrt{1 - \frac{v^2}{c^2}}$$

that tells us that the space *x′* of a reference frame (*x′, t′*), moving with a uniform linear motion at a velocity *v* in respect to a second system of reference (*x, t*), appears to be *contracted* if measured in *(x, t)*; in other words, it appears to be smaller, if measured in the second reference system (*x, t*).

$$m = \frac{m'}{\sqrt{1 - \frac{v^2}{c^2}}}$$

that tells us that the mass *m′* , motionless in a reference frame (*x′, t′*), that is moving with a uniform linear motion at a velocity *v* in respect to a second system of reference *(x, t)*, appears to be *larger* if measured in *(x, t)*; in other words, it appears heavier, if measured in the second reference system (*x, t*).

And finally, the worldwide famous formula:

$$E = m\,c^2$$

that tells us that the energy of an object is strictly related to its mass.

Given the universality meaning of the speed of light *c*, we can affirm that mass and energy are equivalent, are the same thing, after all. And from this fact, it appears evident that from a small amount of mass we can obtain an incredible quantity of energy.

The validity of this first great contribution to theoretical physics by Albert Einstein, called "The Special Theory of Relativity," has been well confirmed by a long series of scientific experiments.

Einstein gave a further, very important contribution to science ten years later, in 1915, with another masterpiece: "The General Theory of Relativity."

# THE GENERAL THEORY OF RELATIVITY

## Newton's gravitation

Let us begin with the gravitational theory introduced by Sir Isaac Newton.

According to tradition, Newton asked himself a question: *if objects, like an apple, fall, does the Moon also fall toward the Earth?*

Newton understood that, in practice, also the Moon must obey the same physical laws as the apples, but he also understood that something else, evidently, is counterbalancing its fall toward the Earth.

From this, only apparently simple question, and from all the other considerations that the famous English scientist made, Newton's laws of gravitation were born. Those laws were—and still are today—able to describe, with a very good precision, the movements of physical bodies inside a gravitational field, celestial objects included.

Planetary orbits around the Sun, as much as the Moon's orbit around the Earth, and the elongated orbits of comets, were no more mysterious.

Now, let us examine the fundamental law of gravitation discovered by Newton.

Newton understood that two masses $m_1$ and $m_2$, placed at a reciprocal distance $r$, are attracted one toward the other with a force $F$ that has a magnitude equal to the product of the two masses, divided by the square of the distance between them, and finally multiplied by a constant $G$, called the Universal Gravitational Constant, or Newton's Constant.

That is to say, in synthesis:

$$F = \frac{m_1 m_2}{r^2} G \tag{1}$$

And the $G$ constant does have a precise, universal value:

$$G = 6.67 * 10^{-11} \, N \, M^2 \, Kg^{-2}$$

Where $N$ is the Newton's Constant, $m$ are meters and obviously $Kg$ are kilograms.

In such a way, the force $F$ is measured in a unit called "Newtons."

From formula *(1)* it derives, obviously, that the attraction force gets bigger when the considered masses get bigger, and the same force gets smaller if the distance between the masses gets bigger.

Let us now consider ourselves on the Earth's surface. Be $m_1$ our mass, and $m_2$ the Earth's mass. And be $r$ the Earth's radius.

If we consider, to a first approximation, that the Earth has a constant radius, we have the following simplification:

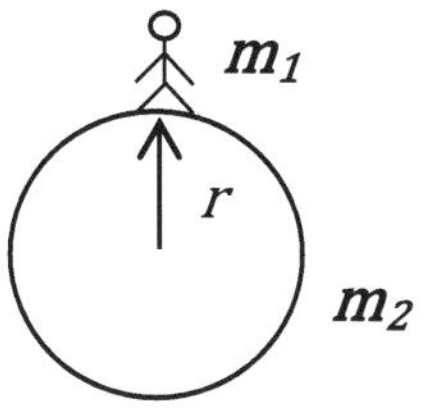


$$F = \frac{m_1 m_2}{r^2} G = g\, m_1$$

having defined:

$$g = \frac{m_2}{r^2} G$$

as the Earth's gravitational acceleration.

Approximately, $g$ = 9.81 m/s$^2$

This means that if my mass is equal to $m_1$, then I undergo an attraction force, exerted by the Earth, equal to:

$$F = g m_1$$

Obviously I do not fall toward the center of the planet Earth, simply because the electromagnetic force that maintains the molecular structure of the surface, on which I am, is able to contrast that gravitational force. But I still find myself in the Earth's action field, and so I can perceive the corresponding gravitational attraction.

Starting from these considerations of a Newtonian kind, the genius of Einstein made gravitational Physics evolve with a very important statement: the "Principle of Equivalence."

# The Principle of Equivalence

Let us imagine to be in an elevator that is free-falling toward the Earth.

The elevator, together with us in its interior, is subjected to the gravitational pull $F = g\ m_1$, being $m_1$ equal to the sum of our mass plus the elevator mass, and being $g$ the acceleration due to the Earth's gravity.

Inside this falling elevator we do not perceive any gravitational force, and so we feel without weight. In fact, we are in a free-fall condition, together with the elevator. No more we can tell, as we have no references other than the elevator itself.

If we had a bead in our hand, and we let it go, in reality it would remain where it was: we would not see it falling toward the floor of our elevator.

In fact, the bead, exactly like us and the elevator, was already in free-fall before we opened the fingers of our hand.

Let us now consider we are in the same elevator, with the same bead in our hand, but this time we are not in a free-fall state toward the Earth's surface. This time we are in deep space, very far from every noticeable mass.

Well, nothing would change in respect to the previous situation: we would not perceive any attraction force, and the bead would not have any motion, relative to us and the elevator, if we let it go.

After those considerations, Einstein stated that an equivalence between the following two situations exists:

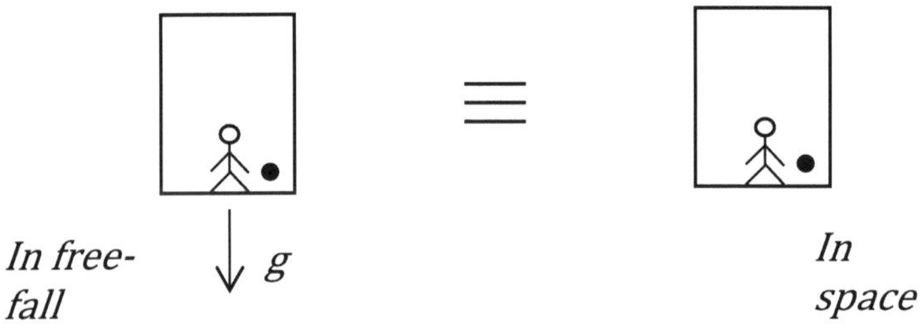


That is to say that a person in an elevator, in free-fall, does not feel any weight and he is in a completely equivalent situation as being in the same elevator placed in space, far away from any mass.

Those two states obviously are not identic, but they are exactly *equivalent* from a physical point of view.

Then Einstein added another consideration: if we are in the same elevator, and this time it rests motionless on the Erath's surface, so that we are subjected to the Earth's gravitational pull $g$, then this situation is equivalent to find ourselves in an elevator that is being pushed by a rocket, with an acceleration equal to $g$ itself:

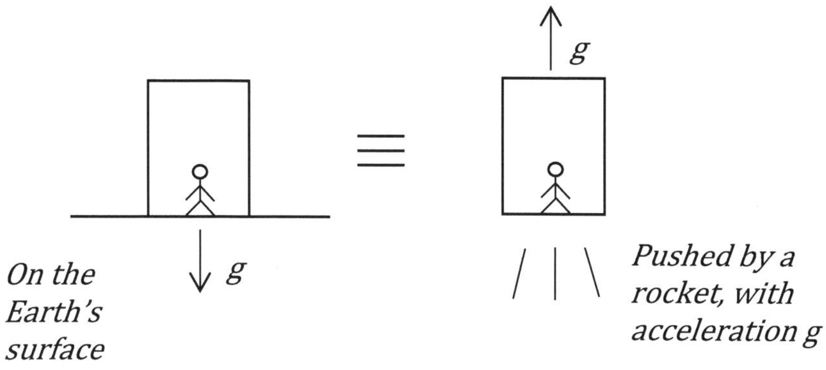


And Einstein also added: there is no experiment, that one can do inside the elevator, that can lead to different results depending on the two situations.

The two physical states are indistinguable the one from the other, staying inside the elevator.

As an example, let us consider this simple experiment: in the two situations (elevator standing still on the Earth and elevator in acceleration, in outer space) let us try and let the bead, that we have in our hand, fall.

In the case of the elevator that is resting on the Earth, the bead obviously will fall down with an accelerated motion, and will reach the elevator floor after a certain amount of time, time that we define $t$.

Let us now consider the situation in which the elevator is being pushed by a rocket in space, with an accelerated motion having an acceleration equal to $g$.

When we open the fingers of our hand, the bead will hit the floor of the elevator after exactly the same amount of time $t$.

In fact, in the moment in which we let it go, the bead had a certain velocity, that it maintained because of the inertial principle, but we and the elevator have continued to accelerate with an acceleration equal to $g$ .

And so, after exactly an elapsed time equal to $t$ , the elevator floor encounters the bead:

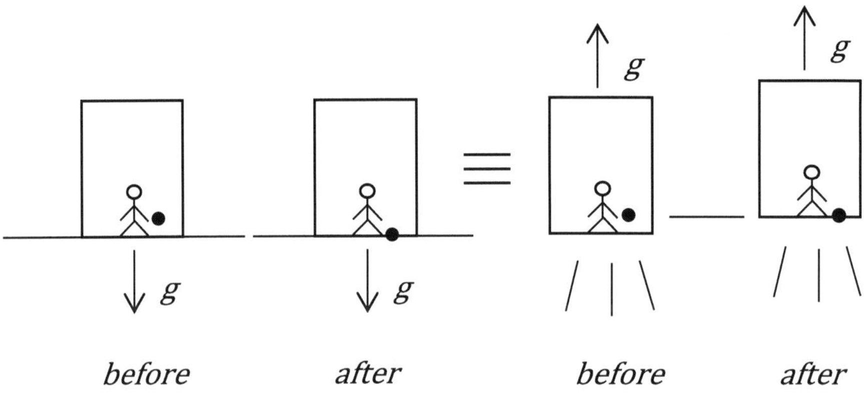


*(please note: in the picture, the bead velocity, at the moment we let it go, is equal to zero only for simplicity of visualization)*

For us, from inside the elevator, there is no way to distinguish between the results of the two experiments: the bead that we have let go seems, in both cases, first falling down with an acceleration $g$ toward the floor of the elevator, and then hitting it after the time $t$ .

The Principle of Equivalence described above is based on the hypothesis that the gravitational field exerted by the Earth is *constant* in every point of space that the elevator occupies, and so, that there is no difference, as an example, between the attraction pull at the level in which our feet are, and at the level in which our head is.

In reality, if we apply Newton's formula (1), we see that our hypothesis is not rigorously true: our feet are at a distance $r$ , from the Earth's center of mass, *shorter* than the distance of our head from the same center of mass.

That is to say that the gravitational force at our feet is slightly higher than the gravitational force that acts on our head.

So we are slightly "stretched" from the gravitational field.

Such an effect is imperceptible on ourselves, because of our limited dimensions, but when the distance $r$ is very high, then this effect is quite evident.

If, for instance, we consider the oceans and the attraction force that the Moon exerts on them, then is quite clear that the water near the equator must be more attracted by the Moon than the water near the poles:

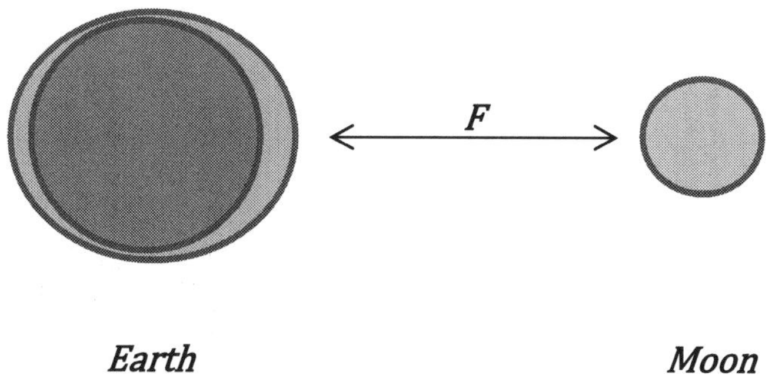

In general, this effect, connected to the variation of the attraction force inside a gravitational field, takes the name of "tidal effect."

In any case, Einstein's Principle of Equivalence is completely correct, if we take by hypothesis that the gravitational field is uniform and constant, at least *locally* (i.e. in the proximity of where the experiment takes place).

# The behavior of light in a gravitational field

Let us now consider again our elevator, imaging it in the situation in which it is pushed by a rocket, with an accelerated motion.

Starting from a point A, placed on a wall of the elevator, a light impulse is emitted.

The photons of that impulse travel toward the opposite wall, while the rocket travels in a uniformly *accelerated* motion.

As we can see on the following picture, because the photons travel on a straight line, at a certain time the impulse is in the point we call B, and B has an inferior distance, from the elevator floor, in respect to the emission point A. After another amount of time, the impulse photons eventually get to the opposite wall of the elevator, hitting it in a point that we define C.

Obviously, the point C is nearer to the floor of the elevator, in respect to the point B.

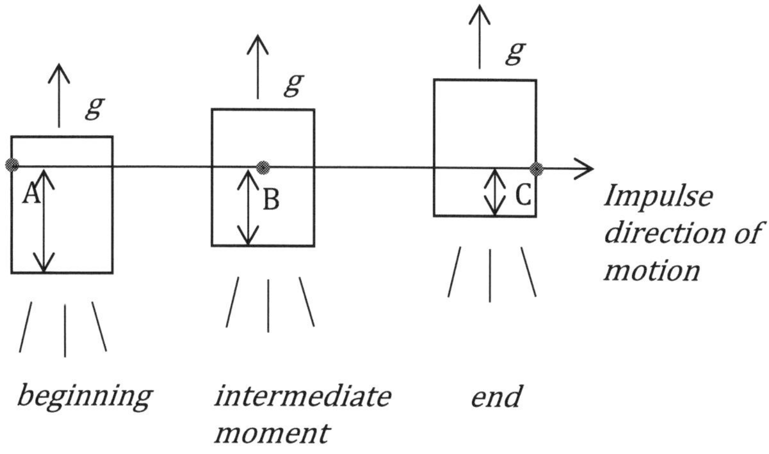

The same phenomenon, as seen from the frame of reference of the elevator (for example as seen by us that are inside the elevator), appears just like the light impulse had a curved trajectory, that bends toward the elevator floor before hitting the arrival wall:

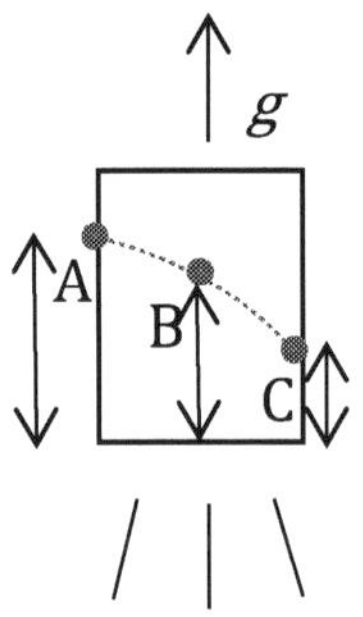


The trajectory appears curved, not straight, because the motion of the elevator is accelerated, that is to say it has not a constant velocity.

Starting from the Principle of Equivalence, Einstein deducted, at this point, that if a person sees a light impulse travel in a curved trajectory when he find himself in an elevator, pushed with an accelerated motion by a rocket, then he would also see exactly the same phenomenon if he found himself in an elevator that is motionless, but placed in a gravitational field with an attraction force equal to the rocket acceleration:

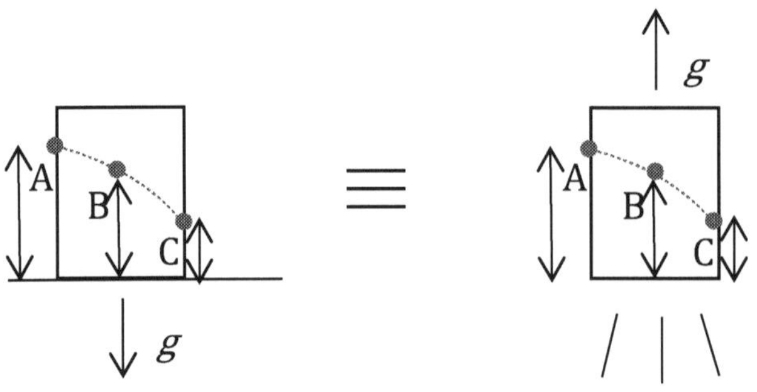

And then, Einstein could affirm that also light, *even though photons do not have mass*, undergoes a deviation in its motion, when it travels in a gravitational field.

This was a *very* revolutionary concept.

# Verification of light curvature in a gravitational field

But how to verify that Einstein was correct, when he asserted that the trajectory of photons bends in a gravitational field?

The magnitudes of the variables, for an experiment to be performed on the Earth (mass of the Earth, speed of light, terrestrial distances, etc.) did not permit to verify that "strange" theory by using the measurement devices present in those times.

But Einstein understood that if his theory was correct, then during a solar eclipse the astronomers would have seen, near the Sun, some stars whose real position was *behind* the Sun.

To have the Sun eclipsed was necessary, in order to avoid that the solar light greatly surpassed the nearby star lights, thus making the measurement impossible.

And exactly what Einstein foresaw, the astronomers did verify: during a solar eclipse they could actually see some stars, whose real position was known, without any doubt, to be *behind* the Sun in that particular day of the year.

What had happened was, that each star image was deflected by the mass of the Sun and each star (in reality in position A, in the picture here below) was seen as if it were in a visible place, not behind the Sun (apparent position B, in the picture here below):

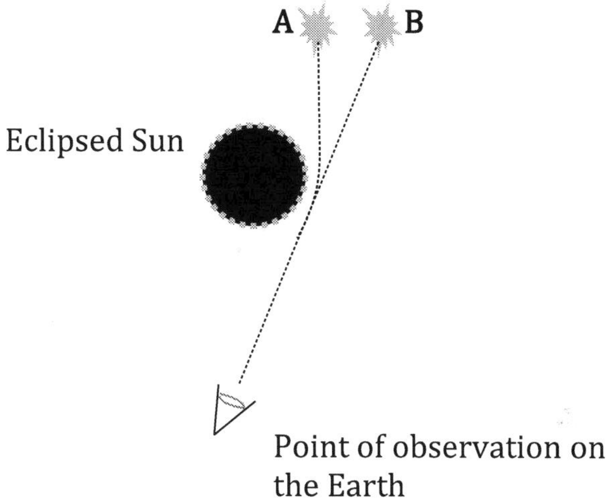


[Simplified scheme of what was really measured by the astrophysicist Arthur Eddington, during the solar eclipse that took place in 1919]

# How does the velocity of light change in a gravitational field?

The measurement of some star images, during a solar eclipse, proved that Einstein's affirmation, that the *trajectory* of light in a gravitational field undergoes a curvature, was correct.

But what else does it happen to light, when it is exposed to a gravitational field?

The speed of light is a universal constant, that in vacuum is equal to:

$$c = 3 * 10^8 \ m/s$$

It is demonstrable, and measurable, that the velocity of light remains constant in modulus (that is, in its magnitude, regardless of its direction), also in a gravitational field.

What really changes is its trajectory, that—as we saw—undergoes a modification in curvature, but not only: also its *frequency* changes, because it undergoes the so called "Doppler effect" (or "Doppler shift").

In fact, let us consider, for simplicity, a light impulse that travels toward the center of mass of a celestial body:

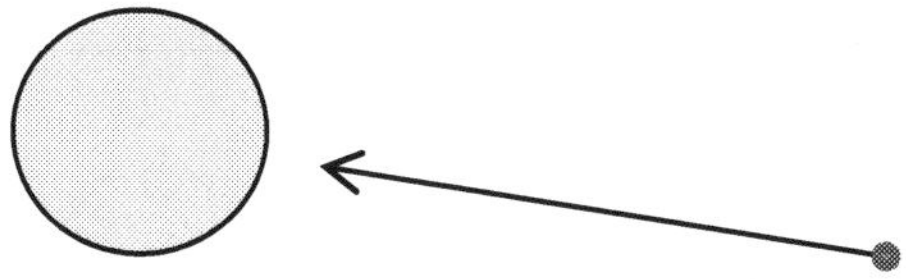

Actually, the photons of the light impulse do not undergo any variation in the magnitude of their velocity, that remains exactly equal to $c$ .

What happens is that the frequency, and so the *color* of that light, changes: as the light gets nearer the celestial body, then its frequency gets higher, and its color shifts toward blue, due to the phenomenon called "blueshift" (if we are speaking about visible light).

Equivalently, if the light impulse travels away from a huge mass, it undergoes a shift of its "electromagnetic spectrum" towards red ("redshift").

Thus, the *vector* of the velocity of light, in a gravitational field, is deflected (it changes direction) and its frequency changes, but its magnitude (or modulus) does not change value.

The explanation of this phenomenon resides again in Einstein's Principle of Equivalence.

In fact, let us consider one more time our elevator, placed on the same rocket with acceleration $g$ . And let us imagine we beam a light impulse from the ceiling of the elevator, downwards.

What happens is—as we can see in the following picture—that the wavelength of the emitted light appears shorter, if measured in the frame of reference of the moving elevator:

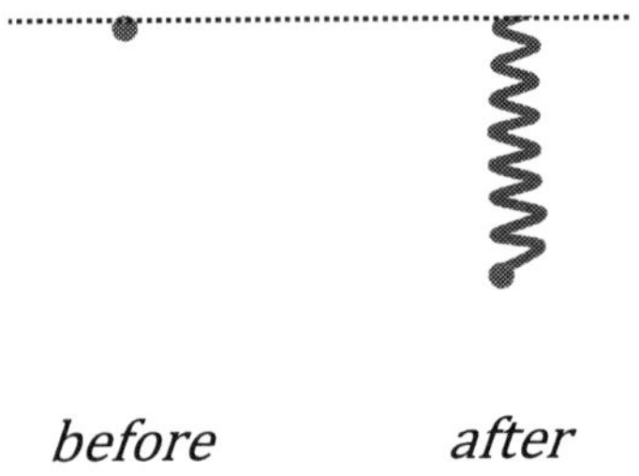


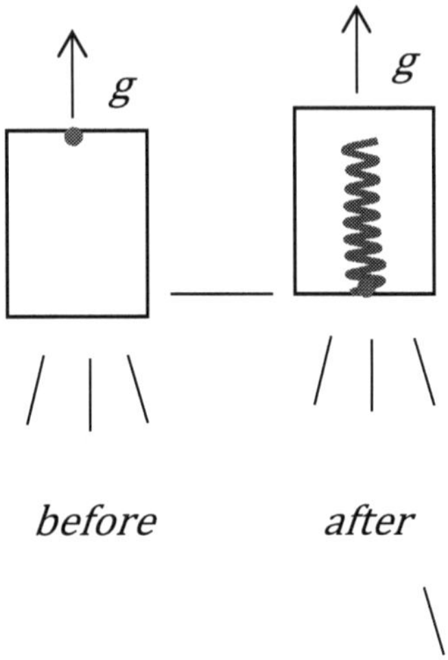


and so its frequency appears to be higher, because:

$$c = \lambda \nu$$

being $\lambda$ the wavelength and $\nu$ the frequency of the emitted light. Thus that light undergoes a blueshift.

Similarly, if we beam the same light impulse from the floor of the elevator, toward the ceiling, then an opposite phenomenon takes place, and the wavelength of the emitted light undergoes a stretch.

And so, its frequency becomes lower, and the light itself undergoes a redshift:

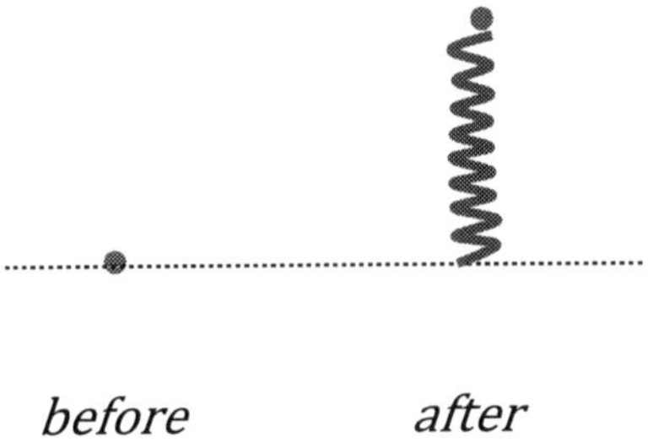


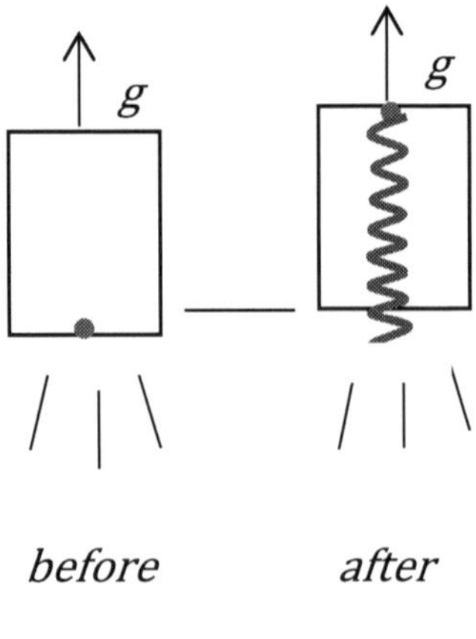

Now, let us apply the Principle of Equivalence: let us consider the same elevator, but this time it is standing still on the Earth's surface. Thus it is subjected to a gravitational field with acceleration $g$ .

Also in this case we would see the same phenomenon. And so, for instance, if a light impulse is emitted from the elevator ceiling toward its floor, the wavelength of the light would appear smaller (and its frequency higher) in respect to a situation in which the elevator is motionless as well, but it is not exposed to a gravitational field:

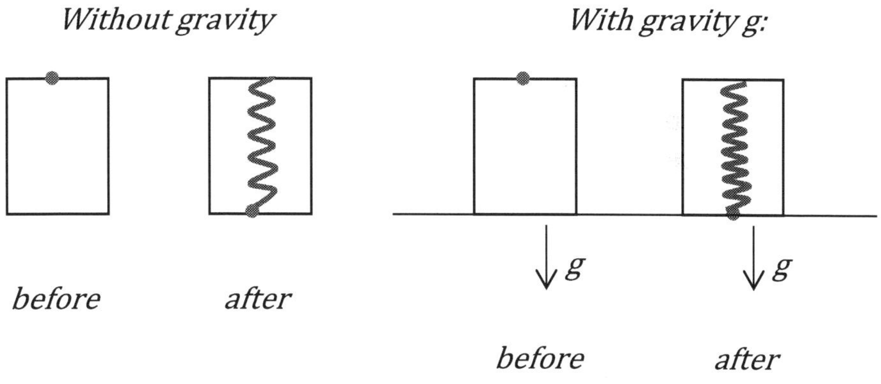

# Space-time curvature

Let us consider a Cartesian frame of reference.

On the ordinate axis we put time, on the abscissa axis we put space.

For simplicity, let us consider only one dimension of space ($x$):

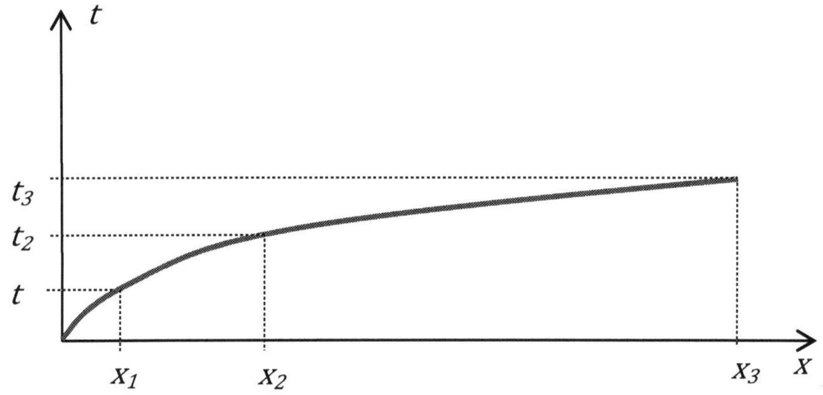


The curve here above represents the motion of our elevator, when it is placed on a rocket that is accelerating in the direction $x$.

In fact, after each same interval of time ($t_1 - 0 = t_2 - t_1 = t_3 - t_2$) the intervals of motion in $x$ are larger and larger. This is by definition, because the motion does not proceed at constant speed, but accelerated.

Such a curve is nothing but the trajectory in space-time (with only one dimension of space and one of time) on which the elevator moves.

By applying again the Principle of Equivalence, we can affirm that space-time, in presence of a gravitational field, undergoes a deformation (curvature) of the kind represented in the picture.

So, if we move near a gravitational field, our local space-time appears modified, and we are not moving on a flat "surface," but on a curved one.

But *what* is responsible of a gravitational field?

Mass. The higher the mass of an object, the more it causes a curvature on space-time.

So we can affirm that **mass**, thanks to the corresponding gravitational field, **"tells" space-time how to curve, and space-time "tells" mass** (but the same is true for objects without mass) **how to move**.

This is the ultimate meaning of Einstein's General Theory of Relativity, obviously far better expressed by the equation system called "Einstein's Field Equation," that here we write for reference, without—by purpose—describing its very difficult physical-mathematical demonstration:

$$R_{\mu\nu} - \frac{1}{2} R g_{\mu\nu} + \Lambda g_{\mu\nu} = \frac{8\pi G}{c^4} T_{\mu\nu} \tag{2}$$

*where:*

$R_{\mu\nu}$= Ricci's curvature tensor

$R$ = scalar curvature

$g_{\mu\nu}$= metric tensor

$\Lambda$ = cosmological constant

$G$ = Newton's gravitational constant

$c$ = speed of light

$T_{\mu\nu}$= stress-energy tensor

# Final considerations

As Einstein himself wrote, his General Theory of Relativity must not be seen as the overcoming of Newtonian theory of gravitation.

Formula *(1)* by Newton is not replaced by Einstein's formula *(2)*.

And in fact, still today, Newton's equations are widely used in a multitude of applications.

Even the difficult trajectory calculation of the space probes, in order to make them reach with great precision very far celestial bodies—like Pluto and many others—are still based on the Newtonian formula of gravitation.

Einstein's theory instead *completes* Newton's huge work, and permits to describe (and predict) the gravitational phenomena in an even more precise way.

Furthermore, considering both the Special and the General Theory of Relativity, we can affirm that the work of such a great scientist has traced the path toward a *deep* comprehension of the universe, that is bringing to mankind a better understanding of how the universe itself evolved starting from the Big Bang, and how it will evolve in the next billions of years.

The relativistic theories are widely considered strong pillars of modern physics.